Alex Dias da Silva

Incidence of fungi in maize seeds (Zea mays L.)

Alex Dias da Silva

Incidence of fungi in maize seeds (Zea mays L.)

Harvest in three micro-regions of Paraíba, Brazil

ScienciaScripts

Imprint
Any brand names and product names mentioned in this book are subject to trademark, brand or patent protection and are trademarks or registered trademarks of their respective holders. The use of brand names, product names, common names, trade names, product descriptions etc. even without a particular marking in this work is in no way to be construed to mean that such names may be regarded as unrestricted in respect of trademark and brand protection legislation and could thus be used by anyone.

Cover image: www.ingimage.com

This book is a translation from the original published under ISBN 978-613-9-71812-2.

Publisher:
Sciencia Scripts
is a trademark of
Dodo Books Indian Ocean Ltd. and OmniScriptum S.R.L publishing group

120 High Road, East Finchley, London, N2 9ED, United Kingdom
Str. Armeneasca 28/1, office 1, Chisinau MD-2012, Republic of Moldova, Europe
Printed at: see last page
ISBN: 978-620-7-95881-8

SUMMARY

The use of high quality seed is extremely important as it prevents the spread of pathogens and the introduction of diseases into areas where they do not exist. Considering the importance of the maize crop for Paraíba, maize (Zea mays L.) seeds were collected during the 2007 and 2008 planting seasons from producing areas in municipalities in the micro-regions of Guarabira (Guarabira, Alagoinha and Mulungu), Brejo (Alagoa Grande and Areia) and Itabaiana (Itabaiana and Gurinhém), in order to determine the mycoflora. The incubation method was moistened filter paper. The following fungi were observed under an optical microscope and the incidence (percentage of affected seeds) was determined: Aspergillus sp., Aspergillus niger, Penicillium spp., Rhizopus sp., Cladosporium sp., Colletotrichum sp., Fusarium spp, Chaetomium sp. and Nigrospora sp. The highest incidence of fungi was recorded in the 2006/2007 harvest with Rhizopus sp. (11.3%) and Fusarium spp. (5.3%) and the lowest incidences of Colletotrichum sp. (0.5%) and Penicillium spp. (1.2%). It can also be seen that the producing micro-regions influenced the occurrence. In the Brejo seeds, the incidence values of the fungi reported were higher than in the other two micro-regions, with Rhizopus sp. (30.0%), Aspergillus sp. (11.125%) and Fusarium spp. (8.75%). In the Itabaiana micro-region, the seeds were the least affected by fungi, with Rhizopus sp. and Fusarium spp. being found, both with an incidence of 1.5%. There were variations in incidence and even occurrence, as was seen with Nigrospora sp. and Chaetomium sp. which were not reported in the 2007/2008 harvest, and Colletotrichum sp. which was absent from the seeds harvested in the 2006/2007 harvest. The mycoflora found on maize seeds in this study was lower in number of fungal species than in another (single) survey carried out in Paraíba. In the Brejo micro-region, rainfall was highest, which, due to the higher relative humidity, may have been the factor behind the higher incidences of fungi found in this micro-region.

Key words: maize, seeds, mycoflora, storage.

DEDICATE

"Those who know your name put their trust in you,
for you, Lord, do not forsake those who seek you."
Psalm 9-10

LIFE

Next thing you know, it's six o'clock!
Next thing you know, it's Friday...
Next thing you know, the year is over...
Next thing you know, 50 years have passed!

Now it's too late to fail... If I was given another chance one day, I wouldn't even look at the clock. I'd just keep going and throwing away the golden, useless shell of the hours along the way... So I say:

Don't stop doing something you like because you don't have time, the only thing you'll miss will be that time, which unfortunately will never come again.

(Mario Quintana)

SUMMARY

CHAPTER 1

INTRODUCTION

The maize crop, Zea mays L., is widespread in our country, and this cereal is planted in regions that differ greatly from one another. In the north-east, maize is important because it is easy to grow and produces well even in precarious technological conditions, and because it is a good source of food for the population and animals. However, this agricultural crop is affected by pests and diseases, from planting to storage, which cause damage to the stalk, leaves, cobs and seeds, reducing or cancelling out production (LUZ, 1995).

According to Pinto (1996), the intensity of the damage caused by pathogens associated with maize seeds depends on the level of seed infection, processing and storage conditions, soil and climate variables and the planting site. In cases of infection, the fungi are usually found as dormant mycelium in the pericarp and endosperm of the seed, which guarantees greater efficiency in survival and transmission to seedlings (CASA et. al, 2005).

Seeds are efficient means of dissemination and transmission and can introduce new pathogens into previously free areas (Rego, 2005). The initial inoculum of the epidemic may depend on the transmission of the pathogen by the seed (Neergaard, 1979). On the other hand, the presence of pathogens can also reduce the physiological quality (germination and vigour) of seeds, which can compromise the establishment of agricultural crops.

According to Machado (1999), the production of pathogen-free seeds is considered the most efficient measure in disease control and is also one of the main factors that condition their performance in the field, the establishment of good stands and, consequently, good productivity (TANAKA, 2001). High seed quality is the basis for increasing agricultural productivity (BORÉM, 2005).

General seed production standards recommend that the material should come from fields free of other species or varieties in order to ensure varietal purity and other quality attributes (Toledo and Marcos Filho, 1977; CESM-PB, 1989). However, in Paraíba's agricultural climate, characterised by the predominance of subsistence crops and the low availability or high cost of certified seeds, farmers produce and store their own seed material.

CHAPTER 2

OBJECTIVES

Considering what has been said in the previous paragraphs, this study was carried out with the following objectives:

2.1. General objective

To determine the mycoflora of seeds produced during the 2006/2007 and 2007/2008 harvests in the micro-regions of Guarabira, Brejo and Itabaiana, in the state of Paraíba.

2.2 Specific objectives

To comparatively assess the incidence of fungi on seeds in the different micro-regions;

Determine the association between the incidence of fungi on seeds and the prevailing climatic conditions (rainfall).

CHAPTER 3

LITERATURE REVIEW

3.1 The socio-economic importance of growing maize

Maize has its origins in the Americas (Central America, Mexico or the south-west of the United States), according to archaeological and geological excavations that show it has been cultivated for at least 5,000 years (Vasconcellos, 1999). It is one of the most widely grown and consumed cereals in the world, whether for human or animal consumption, or as a raw material for various industrial products (FANCELLI and DOURADO NETO, 2000).

The economic importance of maize is characterised by its various uses, ranging from human and animal food to high-tech industry. While the use of maize in the form of grains does not play a significant role in human nutrition, the consumption of its derivatives is an important factor in low-income regions, such as the north-east of Brazil, where maize is the source of energy for many families (MEDEIROS et al., 1998).

As the use of maize for special purposes such as human food increases, the quality of the maize grain becomes more and more important. Grain quality begins with the adoption of a set of practices that must be instituted and monitored until harvest. Failure to control pests that can open doors to diseases in the cobs, the climate and/or other management conditions that are favourable to the development of fungi in the grain can damage it while it is still in the field, affecting its quality and most likely leading to further deterioration during storage (TERRA, 2005).

Maize production doesn't require high technology, but it does require an interest in trying to improve the crop's management conditions more and more, so that producers get better quality production. When monitoring crop conditions, producers should be on the lookout for possible problems in the ear, stalk and root system of the plant. Many of the most common maize diseases are caused by fungi such as Fusarium spp., Diplodia sp., Penicillium spp. and Aspergillus spp., which can attack the stalk and ear through entry doors opened by insects (AGRIANUAL, 2001; TERRA, 2005).

3.2 Sanitary quality of grains and seeds

The use of high quality seeds when establishing maize crops is fundamental for the crop's establishment in the field (Oliveira et al., 1999). As such, analyses of physiological and sanitary quality are essential to ensure the safety of the enterprise. Sanitary analysis is particularly important, as the rigour with which it is carried out prevents seeds from becoming vehicles for pathogens and, consequently, sources of inoculum and the focus of plant epidemics; or from introducing new diseases into areas that are free of them.

The production of high quality seeds is a process that involves knowledge of the principles of integrated disease control, which is applied at all stages of the system: from the field to post-harvest, through the processing stages, the storage period to the final commercialisation stage. At all these stages, a strict quality control scheme must be applied (MACHADO, 1999).

Maintaining the physiological and sanitary quality of maize grains and seeds in general has been the subject of study, mainly due to the prevailing conditions in the places where they are produced and stored (Cruz et al., 1996). The health of maize seeds is one of the main factors conditioning their performance in the field, the establishment of good stands and, consequently, good productivity (Tanaka, 2001). The sanitary quality of the seed is also one of the important aspects to consider when preserving the seed during storage, given the environmental conditions of temperature and humidity (Medina et al., 1995). The loss caused by microflora is not only related to damage to grain quality and quantity, but also to the production of toxic metabolites: mycotoxins (ATHIÉ et al., 1998).

The importance of the occurrence of pathogens in seeds and grains can be assessed from different aspects, all culminating in the economic dimension that each pathogen-host interaction determines. In this way, the economic significance of this association can be estimated based on the expression of each disease as it occurs in nature and considering the fact that most known diseases can have their aetiological agents transmitted effectively by the seeds of their hosts (WEBSTER, 1970).

According to Tanaka et al. (2001), fungal species of the genera Aspergillus, Fusarium and Penicillum occur on maize seeds during and/or after ripening and, depending on the temperature and humidity conditions, can develop and cause damage during storage, reducing the germination and vigour of these seeds and the quality of the grains, a situation that requires the adoption of control measures. Of the Fusarium genus, the F. moniliforme species is a fungus transmitted by maize seeds and grains, causing deterioration, death and toppling of seedlings, rotting of rootlets, stalks, ears and grains.

3.3 Pathogens in maize seeds

Maize seeds can be infected or infested by different fungi during cultivation (field fungi) or in the post-harvest period (storage fungi) and the intensity of the damage caused by these pathogens depends on the level of infection of the seeds, the processing and storage conditions, as well as the soil and climate variables of the planting site (OCHOR et al, 1987; BACON et al., 1994; PINTO, 1996).

Field fungi contaminate grains during cultivation because they require environments with relative humidity above 80%. The survival time of these fungi on seeds is directly related to the conditions of the stored environment (SOUZA, 2007). The most common are: Alternaria, Cladosporium, Fusarium and Helminthosporium (CARVALHO and NAKAGAWA, 2000).

Storage fungi require less water, so they proliferate more intensely in the grain mass in the post-harvest period. In storage, the main fungi are Penicillium spp. and Aspergillus spp. and in the field, Fusarium moniliforme (Richardson, 1979; Goulart, 1994). In addition to these, Cephalosporium sp. is also mentioned by Pinto (1997). Fungi found in crop remains, such as Diplodia maydis, Helminthosporium maydis and Colletotrichum graminicola can aggravate crop damage by rotting both seeds and roots and killing seedlings in pre- and post-emergence (NAZARENO, 1982; PINTO, 1993).

These fungi do not survive competition from other fungi and grow well at high levels of seed moisture content, which is why they do not invade seeds in the field. The fungi that attack stored seeds are called xerophytic and can grow at relative humidity (RH) of up to 70 per cent. These fungi can grow on any organic material that has a moisture content in equilibrium with the relative humidity of the environment of 68 to 90 per cent and are common in nature. The most important fungi in this context are Aspergillus species, the most common of which are A. halophiĺicus, A. restrictus, species from the A. glaucus group (A. amstelodami, A. rubers, A. repens and A. glaucus); A. ochraceus, A. candidus, A. flavus, A. parasiticus Penicillium spp (DHINGRA, 1985).

The main damage caused by the invasion of storage fungi is: loss of germination, discolouration, an increase in the fatty acid content of the grains, degradation of nutritional qualities, combustion and a musty or sour smell and they can show signs of mould or mildew (DHINGRA, 1985; SILVA, 1995; LAZZARI, 1997).

Storage fungi are mainly responsible for the loss of viability of seeds stored with a moisture content above the critical value. It has been shown in various experiments that most of these fungi preferentially attack the embryo, and some only the embryo.

Among the fungi carried by maize seeds, Fusarium *moniliforme* Sheldon stands out due to the frequency and high percentages with which it occurs, both in Brazil and in other countries (Windhan and King, 1983; Casa *et al.*, 1998; Goulart and Fialho, 1999). This fungus is considered to be one of the main causes of seed rot and seedling death, especially when sowing is carried out under stressful conditions, such as high humidity and low temperatures, which are common in the southern region of Brazil during the first plantings in August and September (Lucca Filho, 1987; Casa *et al.*, 1995; Pinto, 1996; Pereira, 1997). In these sub-optimal conditions, the seeds germinate more slowly and are predisposed to infection by the fungus, resulting in low plant populations (Tanaka, 1976; Tanaka and Balmer, 1980; Reis et al., 1995). Since plant population in maize is directly related to yield (Reis et al., 1995), the presence of this pathogen in association with the seeds is considered potentially harmful.

3.4 Studies on maize seed pathology in Paraíba

Some surveys of fungi in maize seeds have already been carried out in the state of Paraíba (Araújo et al, 1984; Araújo et al, 1987; Araújo et al, 2001; Costa and Araújo, 2001), with the most frequent species being: *Aspergillus flavus, A. niger, Aspergillus* spp., *Penicillium* spp., *Cephalosporium* spp., *Nigrospora* sp., *Dreschlera maydis*, *Fusarium moniliforme*, *Borytis* sp., *Candida* sp., *Humicola* sp., *Monilia* sp., *Rhizopus* sp., *Sclerotium rolfsii*, *Stachylidium* sp., *Curvularia* sp., *Phoma* sp., *Rhizoctonia* sp., *Chaetomuim* sp. and *Trichoderma* sp.

Araújo *et al.* (1984) carried out a two-year survey collecting seed samples from farmers' warehouses in different micro-regions of the state and found that the high incidence of fungi, mainly *Penicillium* spp, was responsible for the low germination of the material analysed. The intensity of fungal occurrence (incidence) was affected by the location of storage and the type of packaging used. The highest incidences of storage fungi were found in the municipality of Picuí, in the Curimataú microregion, and for all the microregions the packaging of seeds in metal silos was the most favourable for fungal occurrence, the opposite being true for material preserved in cotton bags.

Regarding the influence of the cultivation system, in the intercropping with black mucuna it was found that doubling the maize plants during the period of maturation and drying of the ears, prior to harvesting, led to a higher incidence of fungi in the seeds (Araújo *et al.*, 2001) of both plants. This cultivation system also affected the mycoflora of the stored seeds, with a high incidence of Penicillium spp, which resulted in reduced germination (COSTA and ARAÚJO, 2001).

CHAPTER 4

MATERIAL AND METHODS

4.1 Seed sampling

During the 2007 and 2008 planting seasons, visits were made to production areas in municipalities in the micro-regions of Guarabira (Guarabira, Alagoinha and Mulungu), Brejo (Alagoa Grande and Areia) and Itabaiana (Itabaiana and Gurinhém), in the state of Paraíba (Figure 1), samples of maize seeds were collected and packed in plastic bags with a capacity of 0.5 kg and taken to the Seed Pathology laboratory at the Agricultural Sciences Centre of the Federal University of Paraíba, Areia/PB, for analysis.

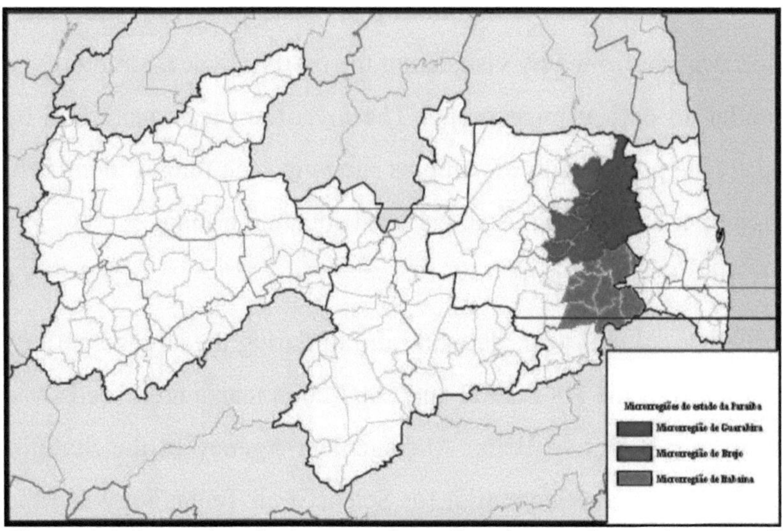

Figure 1: Microregions in the state of Paraíba where maize seeds were collected in the 2006/2007 and 2007/2008 agricultural years.

Source: Google Images (2008).

4.2 Determination of mycoflora

The 200 seeds (four repetitions of 50 seeds) per sample were incubated using the blotter test or moistened filter paper method (Neergaard, 1979). Groups of 10 seeds were placed inside Petri dishes (9.5 cm in diameter), under a substrate made up of two overlapping discs of filter paper moistened with distilled and sterilised water, which was enough to provide the necessary humidity for the development of the possible phytopathogens present in the seeds. The Petri dishes containing the seeds were then taken to a chamber under N. U. lighting. U. V. (Near Ultra Violet) photoperiod of 12 hours and a temperature of 22 ±2 °C. Quantitative and qualitative assessments of the fungi associated with the seeds were made under a stereoscopic microscope on the eighth day of incubation. In some cases, identification was confirmed by visualising the morphological structures of the fungi under an optical microscope. The mycoflora was quantified by considering: 1) the percentages of samples showing each fungus identified; 2) percentages, per sample analysed, per seed containing fungi.

4. 3Climate conditions

For each agricultural year, data on rainfall during the storage periods (August to March) of the 2006/2007 and 2007/2008 maize crop seeds was obtained from the Executive Water Management Agency of the State of Paraíba - AESA, the State Secretariat for Science and Technology and the Environment of the State Government, and also from the offices of the Technical Assistance and Rural Extension Company of Paraíba (EMATER-PB), which can be seen in the following set of figures.

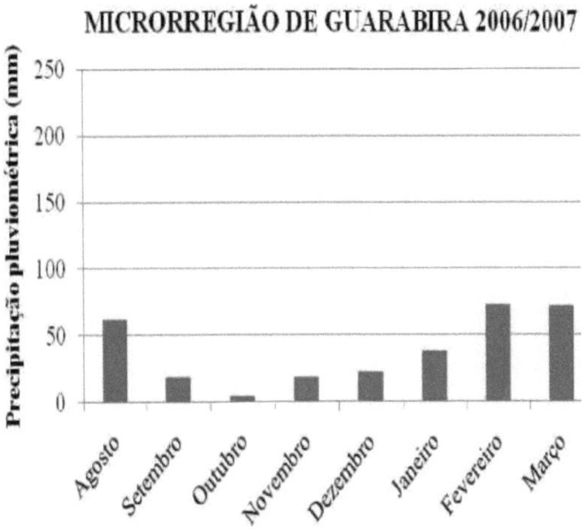

Figura 2: Rainfall recorded in the micro-region of Guarabira, in the state of Paraíba, during the months of August-March in the 2006/2007 agricultural years.

Source: **Own authorship (2008).**

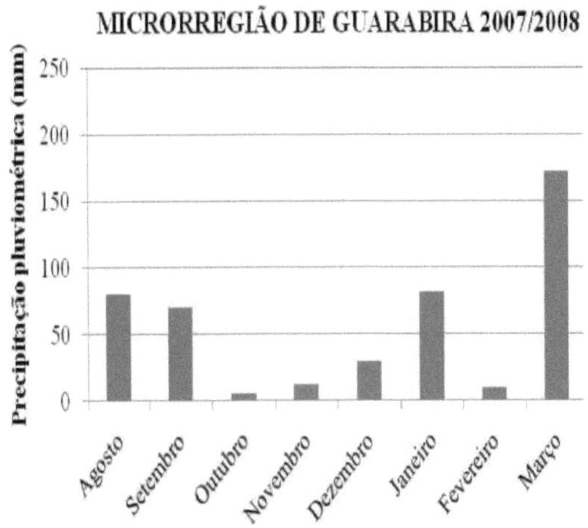

Figura 3: Rainfall recorded in the micro-region of Guarabira, in the state of Paraíba, during the months of August-March in the 2007/2008 agricultural years.

Source: **Own authorship (2008).**

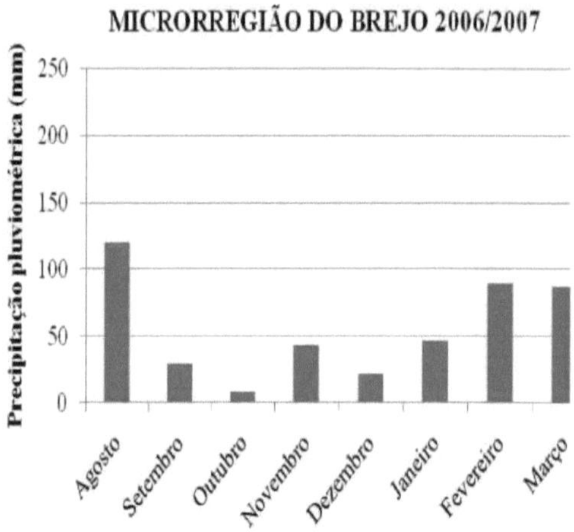

Figura 4: Rainfall recorded in the Brejo micro-region of the state of Paraíba during the months of August-March in the 2006/2007 agricultural years.

Source: **Own authorship (2008).**

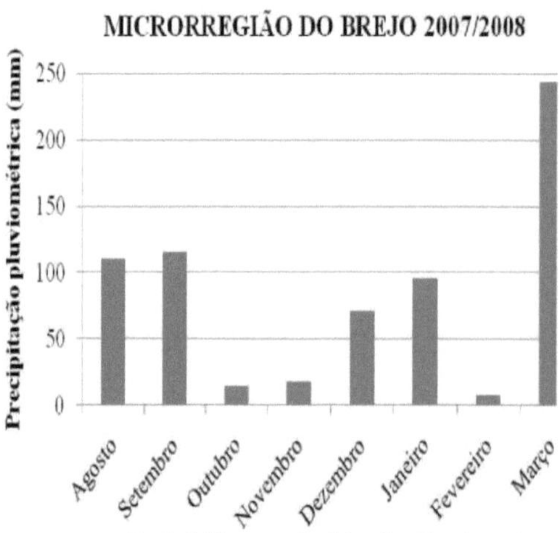

Figura 5: Rainfall recorded in the Brejo micro-region of the state of Paraíba during the months of August-March in the 2007/2008 agricultural years.

Source: **Own authorship (2008).**

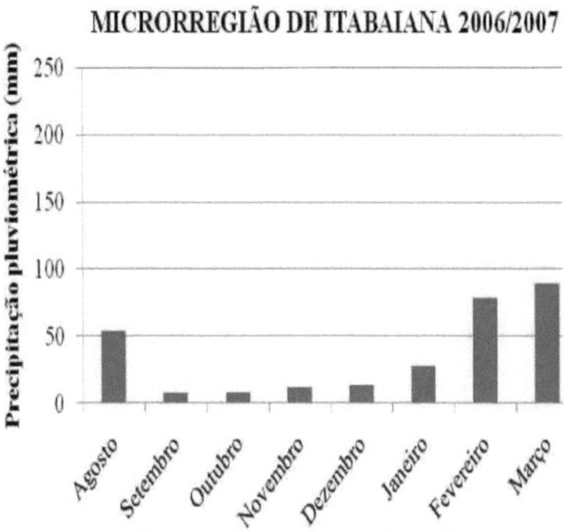

Figura 6: Rainfall recorded in the micro-region of Itabaiana, in the state of Paraíba, during the months of August-March in the 2006/2007 agricultural years.

Source: **Own authorship (2008).**

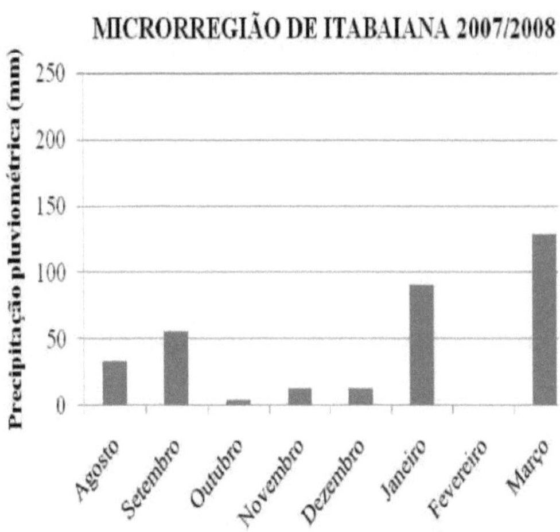

Figura 7: Rainfall recorded in the Brejo micro-region of the state of Paraíba during the months of August-March in the 2007/2008 agricultural years.

Source: **Own authorship (2008).**

CHAPTER 5

RESULTS

5.1 Marked mycoflora

The following fungi were identified in the maize seed samples analysed: Aspergillus sp., Aspergillus niger, Penicillium spp., Rhizopus sp., Cladosporium sp., Colletotrichum sp., Fusarium spp., Chaetomium sp. and Nigrospora sp. whose incidence data is shown in Figures 8 to 13.

5.2 Mycoflora analysis considering the agricultural years

As can be seen in Figure 8, for the average incidence data in the samples analysed for the two periods, the highest values in the 2006/2007 harvest were Rhizopus sp. with 11.3% and Fusarium spp. with 5.3% and the lowest incidences were Colletotrichum sp. with 0.5% and Penicillium spp. with 1.2% No incidences of Nigrospora sp. and Chaetomium sp. were recorded in the seeds collected from the 2006/2007 harvest. In the seeds from the 2007/2008 harvest, the highest incidence values were Fusarium spp. with 4.1% and Rhizopus sp. with 3.9%, and the lowest values were Aspergillus niger and Aspergillus sp., 0.1% and 0.3% respectively. Colletotrichum sp. was not found in the 2007/2008 harvest.

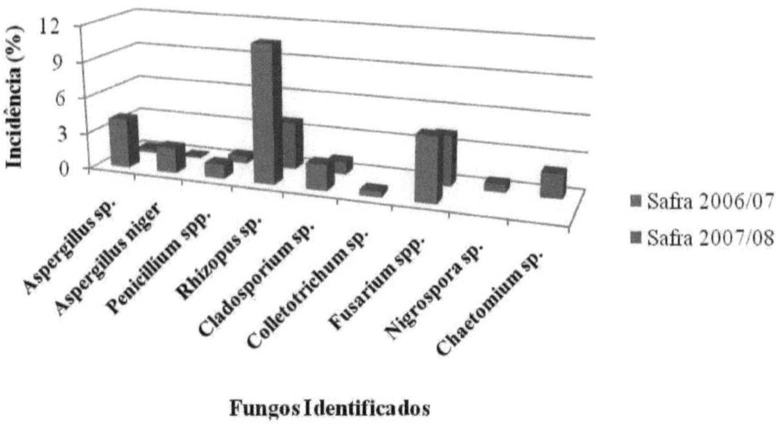

Figure 8: Incidence of fungi - general average - in maize seeds sampled in three micro-regions in the state of Paraíba (Guarabira, Brejo and Itabaiana) in the 2006/2007 and 2007/2008 agricultural years (harvests).

Source: **Own authorship (2008).**

5.3 Analysis of mycoflora considering micro-regions

The results shown in Figure 9 show that in the 2006/2007 harvest, the highest number of fungal species (six) was found in seeds from the Brejo micro-region, although the fungi Colletotrichum sp., Nigrospora sp. and Chaetomium sp. were not reported. The seeds from the Brejo micro-region also recorded the highest incidence rates, in descending order: Rhizopus sp., 30.0%; Aspergillus sp., 11.125%; Fusarium spp., 8.75%; Aspergillus niger, 5.875%; Cladosporium sp., 3.0%, and Penicillium spp., 2.75%. In the Guarabira micro-region, Fusarium spp. was the most common fungus, with 5.625%, followed by Rhizopus sp. with 2.5%; Cladosporium sp. with 2.125%; Aspergillus sp. with 1.125%, and Colletotrichum sp. with 0.625%. In the Itabaiana microregion, the seeds were the least affected by fungi, with Rhizopus sp. and Fusarium spp. being found, both with an incidence of 1.5%.

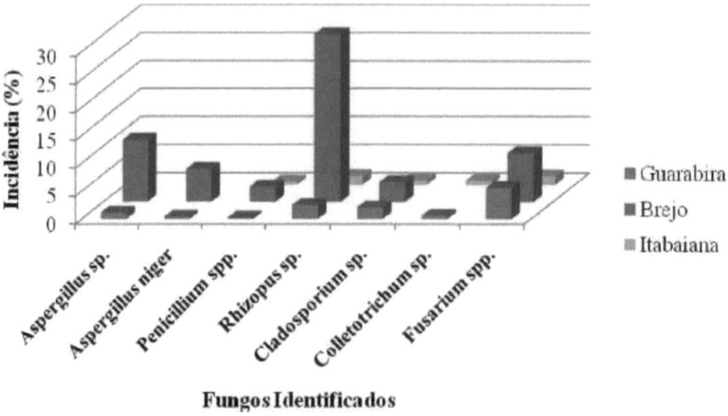

Figure 9: Incidence of fungi - general average - in maize seeds sampled in three micro-regions in the state of Paraíba (Guarabira, Brejo and Itabaiana) in the 2006/2007 agricultural year (harvests).

Source: **Own authorship (2008).**

Still considering the results according to micro-region, Figure 10 shows that in the 2007/2008 harvest the highest number of fungal species (six) was found in seeds from the Brejo micro-region, but in this micro-region the seeds were free of the fungi Aspergillus niger and Chaetomium sp. This year, all the samples were free of Colletotrichum sp. The seeds from the Brejo micro-region also recorded the highest incidence rates, in descending order: Fusarium spp., 8.5%; Rhizopus sp., 4.25%; Cladosporium sp., 1.625%, Penicillium spp., 1.5%; Nigrospora sp., 1.125%, and Aspergillus sp., 0.5%. In the Guarabira micro-region, Rhizopus sp. was the most common fungus, with 4.5%, followed by Fusarium spp. with 3.625%; Chaetomium sp. with 1.25%; Cladosporium sp. with 1.125%; Nigrospora sp. with 0.75%; Aspergillus niger and Penicillium spp. both with 0.125%, and Aspergillus sp. with 0.25%. Seeds from the Itabaiana micro-region were the least affected by fungi and the highest incidences were Chaetomium sp. and Rhizopus sp., with values of 4.75% and 2.875%, respectively.

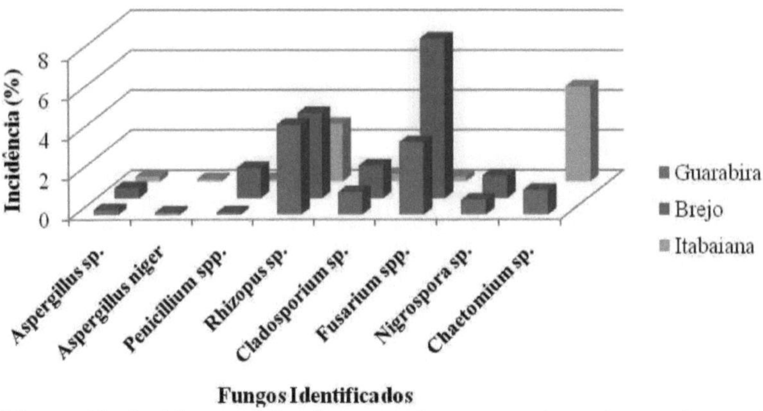

Figure 10: Incidence of fungi - general average - in maize seeds sampled in three micro-regions in the state of Paraíba (Guarabira, Brejo and Itabaiana) in the 2007/2008 agricultural year (harvests).

Source: **Own authorship (2008).**

5. 4Analysis of mycoflora by micro-region according to agricultural years

Evaluating the results from the Guarabira micro-region for the two agricultural years (Figure 11), it can be seen that the highest number of fungal species (eight) was found in seeds from the 2007/2008 harvest. In the seeds from this micro-region collected in the 2006/2007 harvest, the highest incidences were of Fusarium spp. (5.625%) and Rhizopus sp. (2.50%), but there were no incidences of Chaetomium sp. or Nigrospora sp. In the seeds collected in the 2007/2008 harvest, the highest incidences were of Rhizopus sp. (4.50%) and Fusarium spp. (3.625%); Colletotrichum sp. was not observed in these seeds.

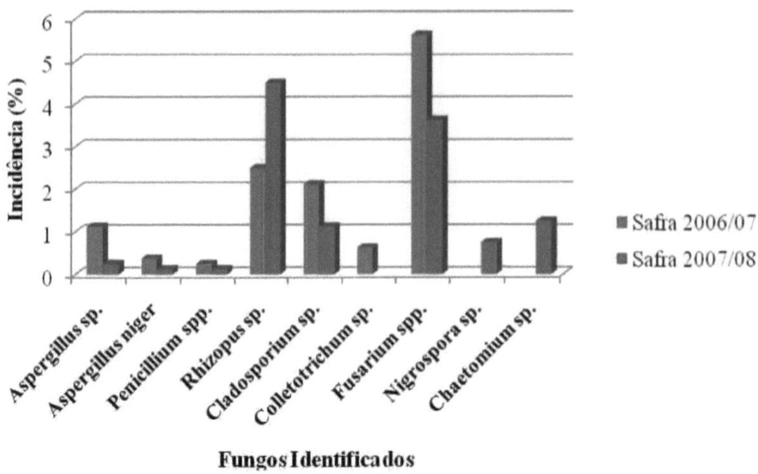

Figure 11: Incidence of fungi - general average - in maize seeds sampled in the Guarabira micro-region, Paraíba state, in the 2006/2007 and 2007/2008 agricultural years (harvests).

Source: **Own authorship (2008).**

In the Brejo micro-region, as can be seen in Figure 12, six fungal species were identified in each agricultural year. In these years, however, all the samples analysed were free of Colletotrichum sp. and Chaetomium sp. Among the fungi reported, Nigrospora sp. was not found in the samples obtained in 2006/2007, and Aspergillus niger was absent from the 2007/2008 seeds. The highest incidence recorded was in the 2006/2007 harvest for the fungus Rhizopus sp. with 30.0%, while the lowest was the incidence of Aspergillus niger (0.5%) in the 2007/2008 harvest.

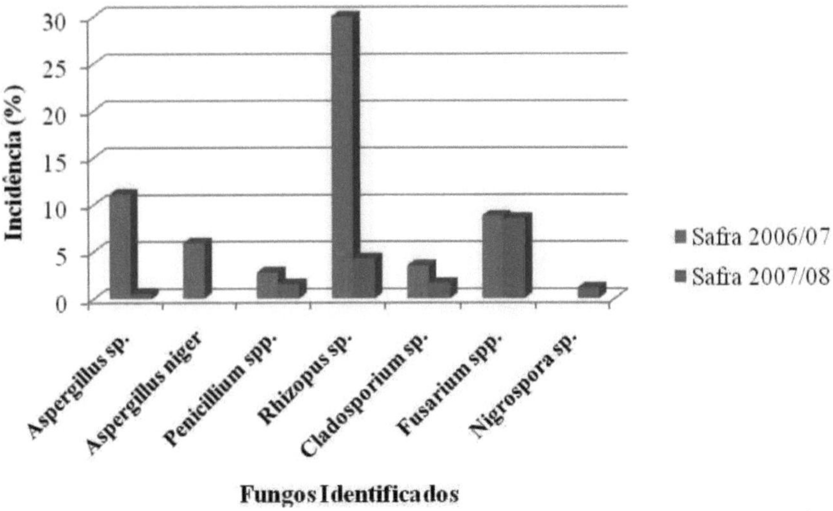

Figure 12: Incidence of fungi - general average - in maize seeds sampled in the Brejo micro-region, state of Paraíba in the 2006/2007 and 2007/2008 agricultural years (harvests).

Source: **Own authorship (2008).**

In the Itabaiana micro-region (Figure 13), the highest number of fungal species (seven) was found in seeds from the 2007/2008 harvest, with the highest incidences in that year being found in the Chaetomium sp. and Rhizopus sp. fungi, with 4.75% and 2.875%, respectively.

Seeds from the 2006/2007 harvest did not contain Aspergillus sp., Aspergillus niger and Chaetomium sp. and the 2007/2008 harvest did not contain the fungus Colletotrichum sp.

Storage fungi (Aspergillus spp. and Penicillium spp.) and Cladosporium sp. and Colletotrichum sp., when recorded, the incidence values were less than 1.0%.

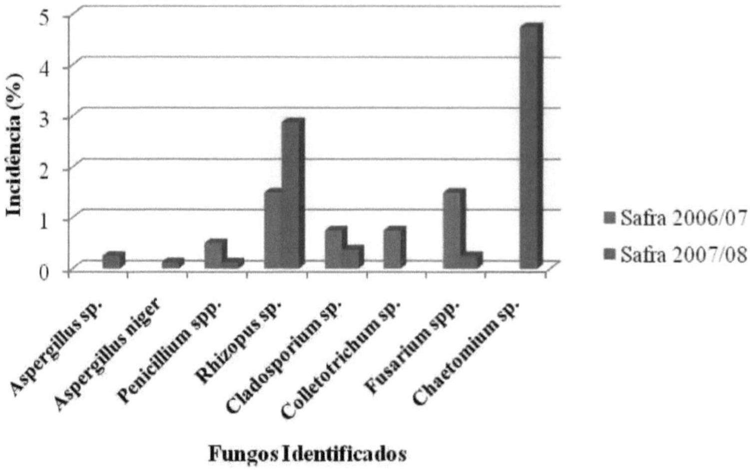

Figure 13: Incidence of fungi - general average - in maize seeds sampled in the Itabaiana micro-region, state of Paraíba in the 2006/2007 and 2007/2008 agricultural years (harvests).

Source: **Own authorship (2008).**

CHAPTER 6

DISCUSSION

The fungi Aspergillus sp., Aspergillus niger, Penicillium spp., Rhizopus sp., Cladosporium sp., Colletotrichum sp., Fusarium spp., Chaetomium sp. and Nigrospora sp. found in this study on maize seeds have already been found in previous studies carried out in Paraíba (Araújo et. al, 1984; Araújo et. al, 1987; Araújo et. al, 2001; Costa and Araújo, 2001). However, in addition to these, Aspergillus flavus, Cephalosporium sp., Dreschlera maydis, Fusarium moniliforme, Borytis sp., Candida sp., Humicola sp., Monilia sp., Sclerotium rolfsii, Stachylidium sp., Curvularia sp., Phoma sp., Rhizoctonia sp. and Trichoderma sp. have also been reported, and Trichoderma sp. On the other hand, those studies did not find Cladosporium sp. and Colletotrichum sp., fungi that were reported in this study and, in the studies carried out by Carneiro (1996) and Soares et al. (2004), were the most frequent on maize seeds and grains.

Among the fungi identified in this study, the genera Aspergillus and Penicillium are considered to be storage fungi and, as well as deteriorating grains and seeds, they also produce mycotoxins (Dhingra and Netto, 1998; Popinigis, 1985; Terra, 2005). The presence of Rhizopus sp., Chaetomium sp. and Nigrospora sp., considered contaminants (Malone and Musket, 1964) corroborates the fact that the material analysed is of poor sanitary quality. Fungi of the genera Cladosporium, Colletotrichum and Fusarium spp. are pathogens of various crops (Souza, 2007). It is worth noting that Fusarium spp. on maize seeds and grains cause a reduction in germination capacity, discolouration or the formation of spots, mould, biochemical transformations and losses in grain and seed weight (Dhingra, 1985; Lucca Filho, 1987). *Colletotrichum* sp. fungi are detected less frequently in maize seeds.

The mycoflora found on maize seeds in this study was lower in number of fungal species than in another (unique) survey carried out in Paraíba (Araújo *et al.* 1984). Two decades have passed since that study was carried out, and the findings of the present study indicate an improvement in the sanitary quality of maize seeds produced and used in the state. This may be due to the introduction of "improved" seeds distributed by government assistance programmes for producers, or that they now have a higher level of technological awareness or more actions resulting from Rural Extension.

As far as storage fungi are concerned, most of the samples analysed had incidence rates of less than 3.0%. In survey work (Araujo *et al.*, 1984) on storage conditions (Araújo *et. al*, 1987; Araújo *et. al*, 2001; Costa and Araújo, 2001), the incidence of *Aspergillus* species ranged from 0.5% to 3.0%, but the records for *Penicillium* ranged from 5.0 to 85.0%. In this study, the incidence of *Penicillium* spp. ranged from 0.5 to 2.5%.

Of the fungi considered to be seed contaminants, such as *Rhizopus* sp., *Chaetomium* sp. and *Nigrospora* sp., the first stands out as it had the highest incidence and was found in all the samples analysed.

For the field fungi, *Cladosporium* sp. and *Fusarium* spp., the former showed incidence values of less than 3.0%, while the latter showed values of more than 5.0%. These results are close to those of studies carried out in Paraíba. For these fungi, the former is of potential importance, as several species in this genus cause diseases in cultivated plants, although no reference was found to diseases they cause in maize. Of the Fusarium genus, some species are important for maize because they affect the grains and F. proliferatum, which is transmitted by seeds, causes toppling (SOUZA, 2007).

Considering the effect of the agricultural years on the fungi reported in this study, there were variations in incidence and even in their presence, as seen with Nigrospora sp. and Chaetomium sp. not reported in the 2007/2008 harvest, and Colletotrichum sp. absent in the seeds harvested in the 2006/2007 harvest.

It can also be seen that the producing micro-regions influenced the occurrence. In the Brejo region, the incidence of the fungus was higher than in the other two micro-regions. In the Brejo micro-region, in the months when rainfall data was recorded, the values were also the highest (Figure 2). In this situation, it can be deduced that the relative humidity, which is an essential climatic variable for the development of fungi, may also have been higher in the Brejo micro-region.

CHAPTER 7

CONCLUSIONS

- The fungi mentioned in this study have already been mentioned in previous studies in Paraíba, with the exception of Cladosporium sp. and Colletotrichum sp;

- The mycoflora found on maize seeds in this study was lower in number of fungal species than in another (single) survey carried out in Paraíba;

- Some fungi found previously were absent in this study;

- There were differences in the incidence of fungi according to the micro-regions and agricultural years in which the seeds were produced;

- In the Brejo seeds, the incidence values of the fungi reported were higher than in the other two micro-regions, while the Itabaiana micro-region was the least affected by fungi.

REFERENCES

AGRIANUAL. Yearbook of Brazilian Agriculture. FNP Consultoria e comércio, 2001, 545p.

ARAÚJO, E.; DORNELAS, G. V.; LIMA, A. L de; BRUNO, R. L. A; OLIVEIRA FILHO, J. J. Evaluation of the quality of maize seeds used for planting in the state of Paraíba. III - maize (Zea *mays* L.) and rice

(Oryza sativa L.). Agropecuária Técnica. Areia, PB, v.5, n. 1/2, p.14-26, 1984.

ARAÚJO, E.; BRUNO, R. L. A.; LIMA, A. L. de DORNELAS, G. V.; BOSCO, J. Influence of different types of packaging and storage environments on the quality of maize seeds (Zea mays L.) in the state of Paraíba. Agropecuária Técnica. Areia, PB, v.8, n. 1/2, p.16-25, 1987.

ARAÚJO, E.; SOUZA, M. A. B. de.; COSTA, R. F. da. Mycoflora of maize and black mucuna seeds produced in isolated and intercropping systems. Agropecuária Técnica. Areia, PB, v.22, n. 1/2, p.59-62, 2001.

ATHIÉ, I.; CASTRO, M. F. P. M.; GOMES, R. A. R.; VALENTINI, S. R. T. Grain conservation. Campinas, Cargill Foundation, 1998. 236p.

BACON, C.W., HINTON, D.M. & RICHARDSON, M.D. A corn seedling assay for resistance to Fusarium moniliforme. Plant Disease Reporter, v.78, p.302-305. 1994.

BORÉM, A. Biotechnology and Seeds. In: ZAMBOLIM, L. Seeds: phytosanitary quality. Viçosa: UFV; DFP, 2005. chap. 1, p. 1-34.

CARNEIRO, J. S. Mycoflora associated with seeds. Fitopatologia

Brasileira, Brasília, v.11, n.2, p.557-567, 1996.

CARVALHO, N. M.; NAKAGAWA, J. Sementes: ciência, tecnologia e produção. 3 ed. Campinas: Cargill Foundation. 2000. 424p

CASA, R. T., REIS, E. M., MEDEIROS, C. A.; MOURA, F. B. Efeito do tratamento de sementes de milho com fungicidas, na proteção de fungos do solo, no Rio Grande do Sul. Fitopatologia Brasileira, v.20, p.633-637, 1995.

CASA, R. T., REIS, E. M.; ZAMBOLIM, L. Fungi associated with maize seed produced in the south and southeast regions of Brazil. Fitopatologia Brasileira, v.23, p.370-373, 1998.

CASA, R. T.; REIS, E. M.; MOREIRA, E. N. Transmission of fungi in winter cereal and maize seeds: epidemiological implications.

In: ZAMBOLIM, L. Seeds: phytosanitary quality. Viçosa: UFV; DFP, 2005. chap. 3, p. 55-74.

CESM-PB, State Seed and Seedling Commission. Technical standards for seed and seedling production. João Pessoa: Ministry of Agriculture/Federal Department of Agriculture, 1989. 85p.

COSTA, R. F. da.; ARAÚJO, E. Influence of cultivation systems and processing on the quality of stored maize seeds (Zea mays L.). Agropecuária Técnica, Areia, PB, v.22, n.1/2, p.27-32, 2001.

CRUZ, A. A.; CRUZ, R. C. B.; CECHINEL FILHO, V.; PINHEIRO, T. R.; ALVEZ JUNIOR, D.; NUNES, R. J.; YUNES, R. A. Antimicrobial effects of N-phenylalkyldichloromaleimides: Structure activity relationship. In: SYMPOSIUM OF MEDICINAL PLANTS OF BRAZIL, 14. Abstracts. Florianópolis, 1996.

DHINGRA, O. D. Damage caused by microorganisms during seed storage. Revista Brasileira de Sementes, 7:139-144. 1985.

DHINGRA, O. D. and NETTO, R. A. C. Mycotoxins in grains. RAPP, v.6, p.49-101, 1998.

FANCELLI, A. L.; DOURADO NETO, D. Maize production. Guaíba: Agropecuária, 2000, 360p.

GOULART, A. C. P. Sanitary quality of "BR-201" maize seeds in the Dourados-MS region. ABRATES Newsletter, 4:53-5, 1994.

GOULART, A. C. P. and FIALHO, W. F. B. Incidence and control of Fusarium moniliforme in maize seeds. ABRATES Newsletter 9:110,

1999.

LAZZARI, F. A. Humidity, fungi and mycotoxins in the quality of seeds, grains and feed. Curitiba: Ed. do Autor, 1997. 140p.

LUCA FILHO, O. A. Maize seed health tests. In: SOAVE, J. & WETZEL, M.M.V.S. (Ed.) Seed pathology. Campinas: Cargill Foundation, 1987. p.430-440.

LUZ, W. C. da. Illustrated systematic key to the main maize seed fungi. Technical Circular No. 7, EMBRAPA CNPT, Passo Fundo, 28p., 1995.

MACHADO, J. C. Seed health management in disease control. Lavras: UFLA/FAEPE, 1999. 78p.

MALONE, J. P. and MUSKET, A. C. Seed borne-fungi: description of 77 fungus species. Proceedings of the International Seed Testing Association, v.29, n.2, p.174-384, 1964.

MEDEIROS, J. F.; PEIXOTO, J. F. S.; GUEDES, J. M. R.; SILVA, I. F.; ANDRADE, A. P. Crop residue management systems in maize (Zea mays L.): ear insertion height and plant height. IN: UFPB Scientific Initiation Meeting, 6. Abstracts... João Pessoa, 1998.

MEDINA, P. F.; RAZERA, F. L.; ROSSETO, C. J. Storage of seeds treated with insecticides and fungicides. Revista Brasileira de Sementes, Brasília, v.17, n.2, p.236-242, 1995.

NAZARENO, N. R. X. Disease control. In: Iapar. Maize in Paraná. Circular Iapar, 29:149-63, 1982.

NEERGAARD, P. Seed pathology. 2.ed. London: Mac Millan Press, 1979. v.2, 1191p.

OCHOR, T. E., TREVATHAN, L. E.; KING, S. B. Relationship of harvest date and host genotype to infection of maize kernels by Fusarium moniliforme. Plant Disease Reporter, v.71, p.311-313, 1987.

OLIVEIRA, W. F.; MIRANDA, B. A.; CAETANO, F. V. C.; MACHADO, L. A. and RAMALHO, V. Effect of phytosanitary products in the treatment of maize seeds (Zea mays L.), aiming at the control of Fusarium moniliforme (Sheld). Pesquisa Agropecuária Brasileira, 29(1): 65-69, 1999.

PEREIRA, O. A. P. Maize diseases. In: KIMATI, H., AMORIM, L., BERGAMIN FILHO, A., CAMARGO, L. E. A. & REZENDE, J. A. M. (Ed.) Manual de Fitopatologia. São Paulo. Editora Agronômica Ceres, p.538-555, 1997.

PINTO, N. F. J. A. Seed treatment with fungicides. In: Technology for maize seed production. Sete Lagoas: EMBRAPA/CNPMS, 1993. p.43-47 (Technical Circular, 19).

PINTO, N. F. J.A. Maize seed treatment. BRAZILIAN SYMPOSIUM ON SEED PATHOLOGY, 4. Proceedings... Gramado, RS, Cargill Foundation, p.52-57, 1996.

PINTO, N. F. J. A. Sanitary analysis in the production of seeds of major crops. In: ZAMBOLIM, L. Seeds: phytosanitary quality. Viçosa: UFV; DFP, 2005. chap. 11, p. 295-332.

POPINIGIS, F. Seed physiology. Brasilia, AGIPLAN, 1985. 289p.

REGO, A. M. Phytosanitary analysis in vegetable seed production. In: ZAMBOLIM, L. Seeds: phytosanitary quality. Viçosa: UFV; DFP, 2005. chap. 10, p. 267-294.

REIS, A. C.; REIS, E. M.; CASA, R. T.; FORCELINI, C. A. Eradication of pathogenic fungi associated with maize seeds and protection against Pythium sp. present in the soil by fungicide treatment. Fitopatologia Brasileira, v.20, p. 585-590, 1995.

RICHARDSON, M. J. 1979. An annotated list of seedborne diseases. 3.ed.

Zurich. CAB/CMI/ISTA. Phytopathological Papers, 23. 320p.

SILVA, J. S. Pre-processing of agricultural products. Juiz de Fora: Maria Institute, 1995. 509p.

SOARES, C. S.; BRUNO, R. L. A.; SILVA, I. F.; BRUNO, G. B. Mycoflora of maize seeds (Zea mays L.) produced in different cultivation systems. VIII BRAZILIAN SEED PATHOLOGY SYMPOSIUM. João Pessoa-PB, 12 to 15 Oct 2004.

SOUZA, A. E. R. Antimicrobial activity of garlic and santo grass extracts on the development of Fusarium moniliforme isolated from maize grains (Master's thesis). Areia: UFPB, 2007, 106p.

TANAKA, M. A. S. Factors influencing the germination of maize seeds (Zea mays L.) in the presence of Fusarium moniliforme Sleldon. (Master's Thesis). Piracicaba. Luiz de Queiroz College of Agriculture, University of São Paulo. 1976.

TANAKA, M. A. S. and BALMER, E. Effect of temperature and associated microorganisms on the germination of maize (Zea mays L.) seeds. Fitopatologia Brasileira, v.5, p.87- 93, 1980.

TANAKA, M. A. S. Survival of Fusarium moniliforme in maize seeds kept under two storage conditions. Fitopatologia Brasileira, v.26, p.60-64, 2001.

TANAKA, M. A. S.; MAEDA, J. A.; PLAZAS, I. H. A. Z. Fungal mycoflora of maize seeds in storage environments. Sci. Agric., July/Sept. 2001, v.58, n.3, p.501-508.

TERRA, B. Maintaining grain quality during harvest. Pioneer Newsletter. Rural Administration. Year IX, n.20, p.10-2, 2005.

TOLEDO, F. F.; MARCO FILHO, J. Manual das sementes. Production Technology. São Paulo: Agronômica Ceres, 1977. 244p.

VASCONCELLOS, J. H. Cereal from the Americas to the world. Folha da EMBRAPA, Brasília, v.7, n.37, p.1-12, 1999.

WEBSTER, J. Introduction to fungi. Cambridge: University Press, 1970, 424p.

WINDHAM, M. T. and KING, S. B. Micoflora of roots of maize at seedling and silking stages in Mississippi. Plant Disease Reporter, v.67, p.1366- 1368, 1983.

MIX
Papier aus verantwortungsvollen Quellen
Paper from responsible sources
FSC® C105338

Printed by Books on Demand GmbH, Norderstedt / Germany